SCHOLASTIC

studySMART

Numbers 1 to 30

Level K2

Mathematics

Joan Novelli & Holly Grundon

Welcome to studySMART !

Numbers 1 to 30 reinforces your child's classroom learning of number recognition, number formation, quantities and counting. It provides the practice your child needs to thoroughly learn the numbers 1 to 30.

A variety of engaging activities for numbers 1 through 30 include:

- Tracing and writing numbers
- Identifying and sequencing numbers
- Drawing specified quantities
- Counting objects in a set
- Comparing sets of objects

The systematic and predictable write-and-learn activities encourage independence and feelings of success. Parent notes at the end of the activities provide ideas for extending your child's learning.

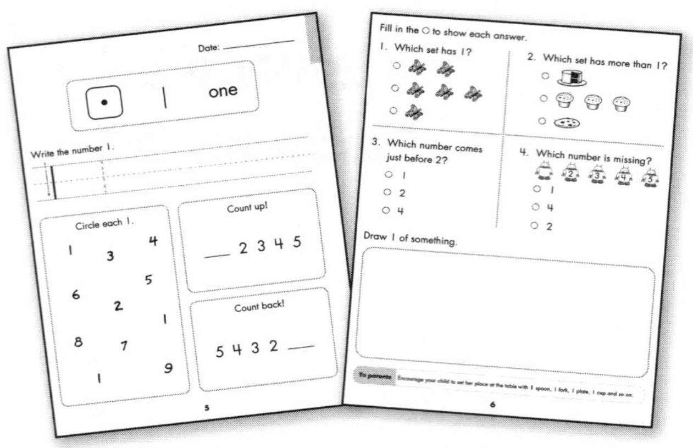

Review pages throughout the book and a number chart to further support learning.

Note: To avoid the awkward 'he or she' construction, the pronouns in this book will refer to the female gender.

Contents

Number Chart

1	2	3	4	5
6	7	8	9	10
11	12	13	14	15
16	17	18	19	20
21	22	23	24	25
26	27	28	29	30

Date: ____________

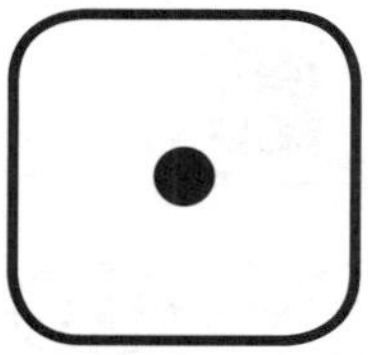

1 one

Write the number 1.

1

Circle each 1.

1 4
3
5
6
2
1
8
7
1 9

Count up!

___ 2 3 4 5

Count back!

5 4 3 2 ___

Fill in the ○ to show each answer.

1. Which set has 1?

○ 

○ 

○ 

2. Which set has more than 1?

○ 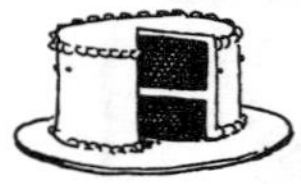

○ 

○ 

3. Which number comes just before 2?

○ 1

○ 2

○ 4

4. Which number is missing?

○ 1

○ 4

○ 2

Draw 1 of something.

To parents Encourage your child to set her place at the table with **1** spoon, **1** fork, **1** plate, **1** cup and so on.

Date: ____________

2 two

Write the number 2.

2 2

Circle each 2.

9 5 2 7 4 3 8 1 2 2 1 6 1

Count up!

1 ___ 3 4 5

Count back!

5 4 3 ___ 1

Fill in the ○ to show each answer.

1. Which set has 2?

○ 

○ 

○ 

2. Which set has more than 2?

○ 

○ 

○ 

3. Which number comes just before 3?

○ 1

○ 2

○ 4

4. Which numbers are missing?

○ 4 and 5

○ 1 and 2

○ 3 and 4

Draw 2 of something.

To parents Go to your shoe closet and have your child count '1, **2**' for each pair of shoes. Y
same activity using socks.

Date: ____________

 3 three

Write the number 3.

3 3

Circle each 3.

6 3 2

5

1 4

3 1

8 9

1 3 7

Count up!

1 2 ____ 4 5

Count back!

5 4 ____ 2 1

Fill in the ○ to show each answer.

1. Which set has 3?

○ 

○ 

○ 

2. Which set has more than 2?

○ 

○ 

○ 

3. Which number comes just before 3?

○ 1

○ 3

○ 2

4. Which numbers are missing?

○ 1 and 4

○ 2 and 3

○ 4 and 5

Draw 3 of something.

To parents Take a button-front shirt and ask your child to do up any **3** buttons. Encourage her to count up to 3 as she buttons her shirt.

Date: ____________

4 four

Write the number 4.

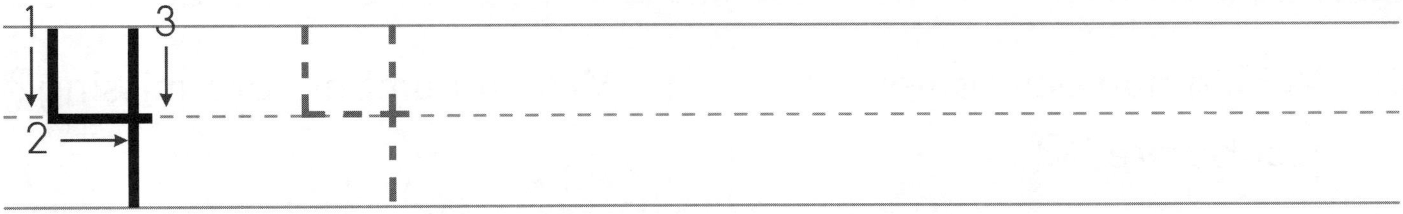

Circle each 4.

5 1 8

6 3

1 2 4

4 1

4 7 9

Count up!

1 2 3 ____ 5

Count back!

5 ____ 3 2 1

Fill in the ○ to show each answer.

1. Which set has 4?

○ 

○ 

○ 

2. Which set has more than 3?

○ 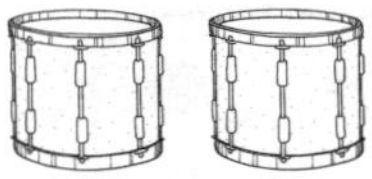

○ 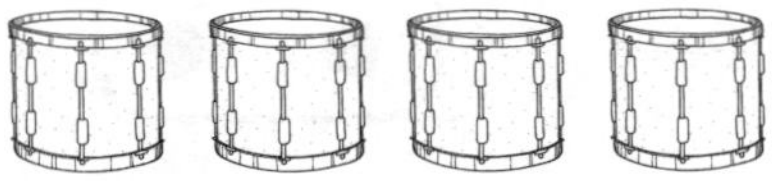

○ 

3. Which number comes just before 4?

○ 2

○ 3

○ 1

4. Which numbers are missing?

○ 1 and 3

○ 3 and 5

○ 2 and 4

Draw 4 of something.

To parents Help your child make a list of things that come in fours — sides of a square or rectangular table, legs of a cat and so on.

Date: ____________

5 five

Write the number 5.

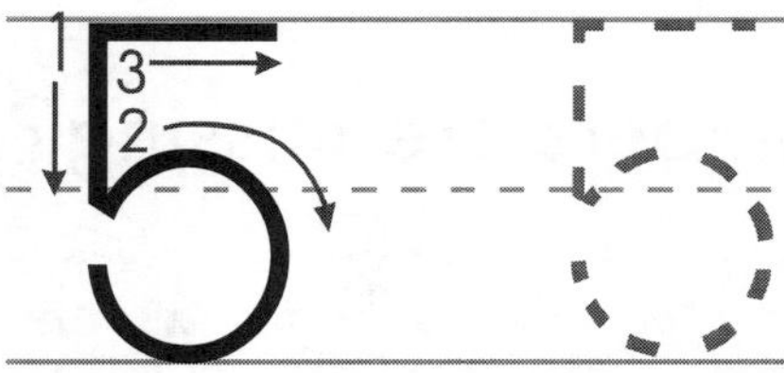

Circle each 5.

1 3 4

6 5

2

8 7

5

1 9

Count up!

1 2 3 4 ____

Count back!

____ 4 3 2 1

Fill in the ○ to show each answer.

1. Which set has 5?

○ 

○ 

○ 

2. Which set has more than 3?

○ 

○ 

○ 

3. Which number comes just before 5?

○ 1

○ 3

○ 4

4. Which numbers are missing?

○ 1 and 4

○ 2 and 5

○ 3 and 5

Draw 5 of something.

To parents Place your child's hand on a piece of chart paper and trace around it with a marker. Count the number of fingers on the tracing, with your child.

Date: ______________

Review Numbers 1 to 5

Circle the number that tells how many.

1 3 5	
2 1 3	
1 4 5	
2 4 1	
2 1 3	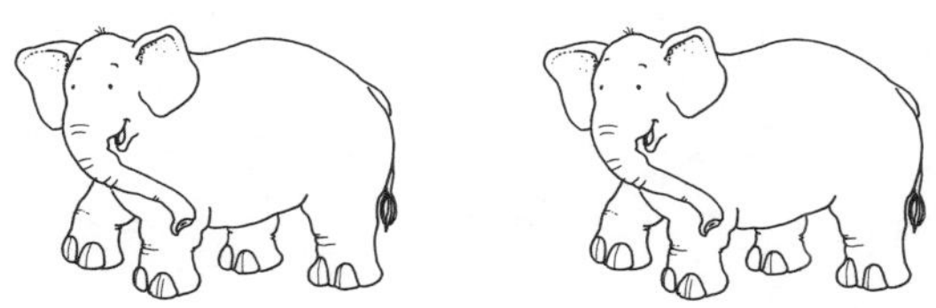

Count and write the number in the box.
Color the set that has **more** balloons.

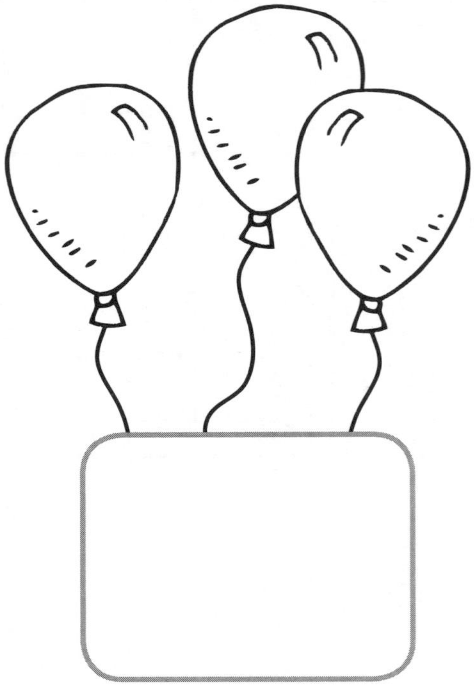

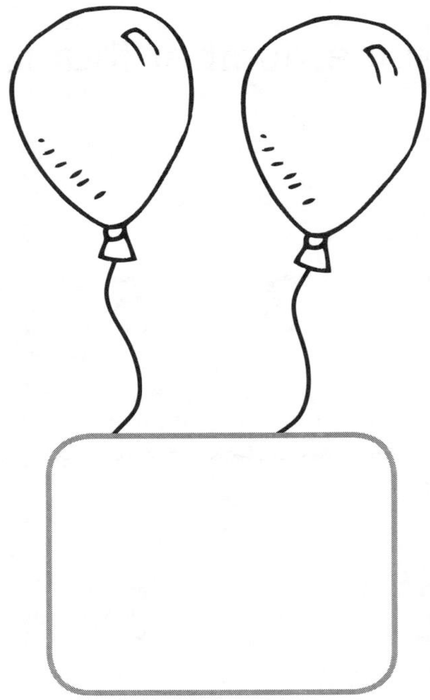

Count and write the number in the box.
Color the set that has **fewer** balloons.

Date: ____________

6 six

Write the number 6.

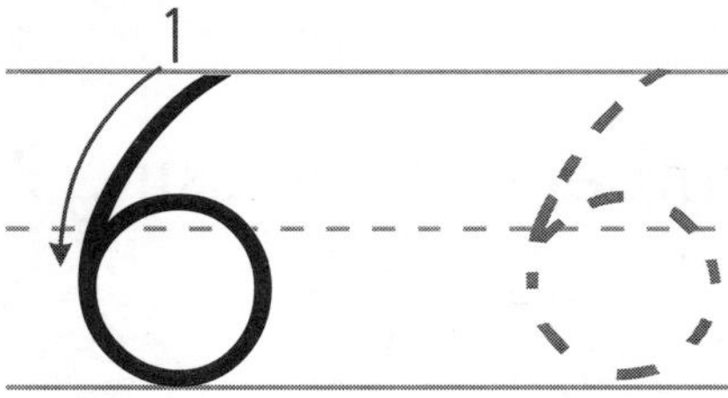

Circle each 6.

1 3 4 5 6 2 7 8 6 1 9

Count up!

2 3 4 5 ____

Count back!

____ 5 4 3 2

Fill in the ○ to show each answer.

1. Which set has 6?

○ 

○ 

○ 

2. Which set has more than 4?

○ 

○ 

○ 

3. Which number comes just before 6?

○ 1

○ 3

○ 5

4. Which numbers are missing?

○ 1 and 2

○ 4 and 5

○ 3 and 6

Draw 6 of something.

To parents At the supermarket, show your child a 6-pack carton of juice boxes. Have her count out the number of juice boxes in the carton. You can also show her a half-dozen egg carton and have her count out the **6** eggs.

Date: ____________

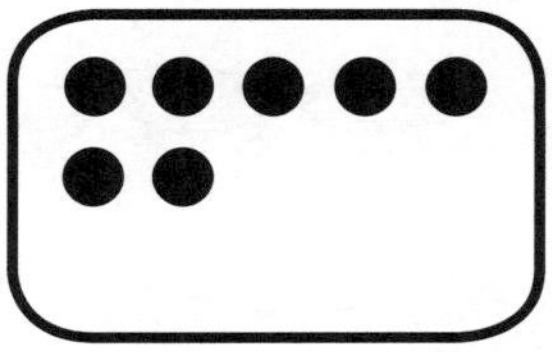

7 seven

Write the number 7.

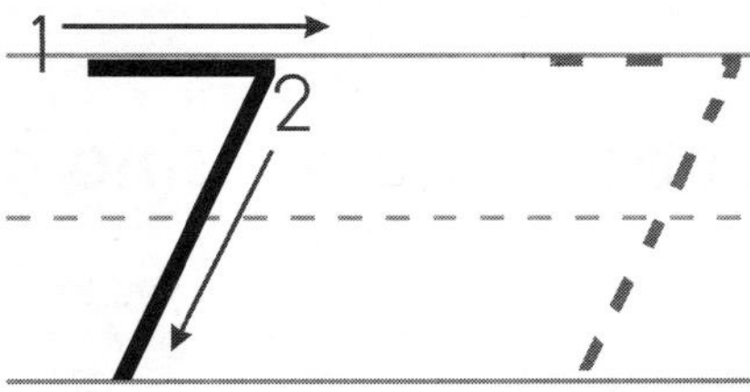

Circle each 7.

6 3 4

7 5

2 7

8 7 1

1 9

Count up!

3 4 5 6 ____

Count back!

____ 6 5 4 3

Fill in the ○ to show each answer.

1. Which set has 7?

○ 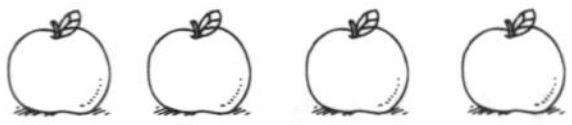

○ 

○ 

2. Which set has more than 5?

○ 

○ 

○ 

3. Which number comes just before 7?

○ 4

○ 6

○ 2

4. Which numbers are missing?

○ 4 and 5

○ 1 and 2

○ 3 and 6

Draw 7 of something.

To parents	Tell your child that there are ***7*** days in a week. Let her practice counting the number of days in each week using a page of the calendar.

Date: ____________

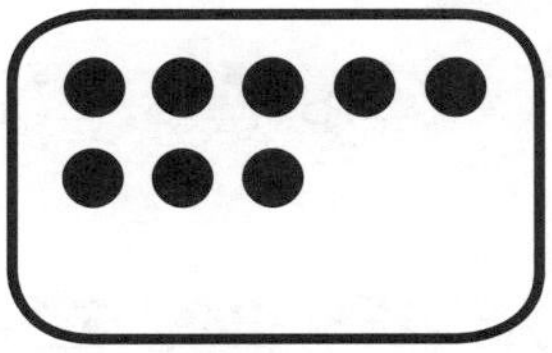

8 eight

Write the number 8.

8 8

Circle each 8.

1 3 4

6 5

2 8

8 7

1 8 9

Count up!

4 5 6 7 ____

Count back!

____ 7 6 5 4

Fill in the ○ to show each answer.

1. Which set has 8?

○ 

○ 

○ 

2. Which set has more than 6?

○ 

○ 

○ 

3. Which number comes just before 8?

○ 7

○ 2

○ 4

4. Which numbers are missing?

○ 1 and 2

○ 6 and 8

○ 2 and 3

Draw 8 of something.

To parents Give your child a box of colored pencils. Have her count and pull out **8** of them from the box.

Date: ______________

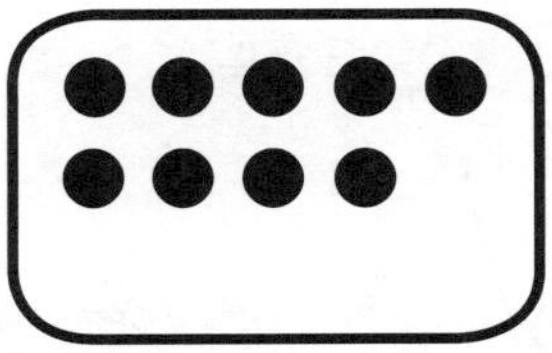

9 nine

Write the number 9.

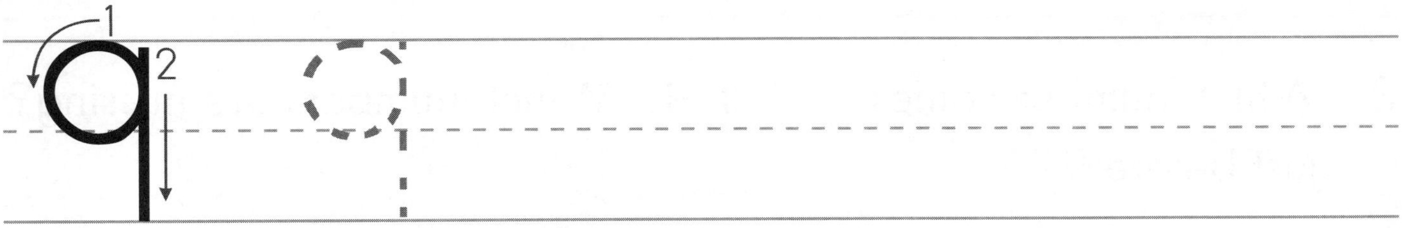

Circle each 9.

9 3 4

6 5

9 2

8 7

1 9

Count up!

5 6 7 8 ____

Count back!

____ 8 7 6 5

Fill in the ○ to show each answer.

1. Which set has 9?

○ 

○ 

○ 

2. Which set has more than 7?

○ 

○ 

○ 

3. Which number comes just before 9?

○ 8

○ 3

○ 1

4. Which numbers are missing?

○ 5 and 6

○ 2 and 3

○ 1 and 4

Draw 9 of something.

To parents Take out all the spoons in your kitchen drawer. Ask your child to count **9** spoons and line them up on a table.

Date: ____________

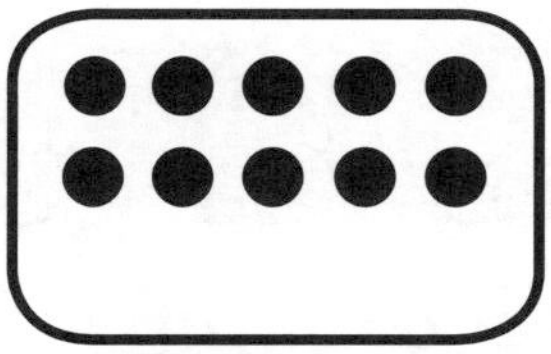 10 ten

Write the number 10.

10 10

Circle each 10.

10 3 4

6 5

2 10

8 7

10 9

Count up!

6 7 8 9 ____

Count back!

____ 9 8 7 6

Fill in the ○ to show each answer.

1. How many ▲?

▲▲▲▲▲▲▲▲▲▲

○ 7

○ 5

○ 10

2. Which numbers are missing?

○ 3 and 4

○ 9 and 10

○ 1 and 2

3. Which number comes just before 10?

○ 3

○ 7

○ 9

4. Which number is the largest?

○ 8

○ 10

○ 1

Draw 10 of something.

To parents Sit by a bookshelf at home with your child. Ask her to count and pull out **10** books.

Date: ______________

Review Numbers 6 to 10

Circle the number that tells how many.

4 6 2	
5 10 1	
9 3 8	
5 7 8	
7 5 1	

Count and write the number in the box.
Color the set that has **more** balloons.

Count and write the number in the box.
Color the set that has **fewer** balloons.

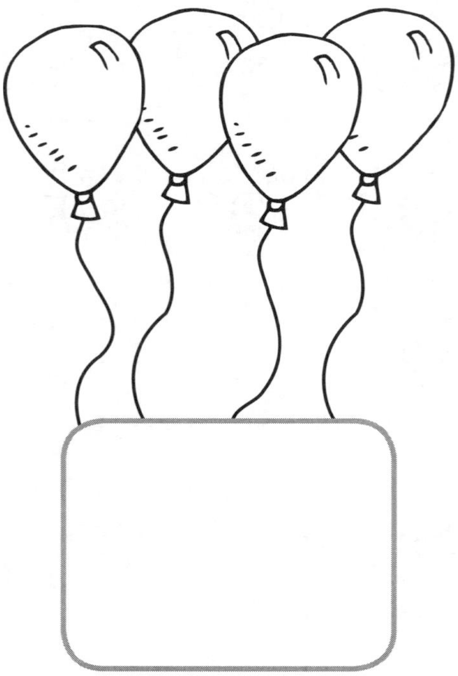

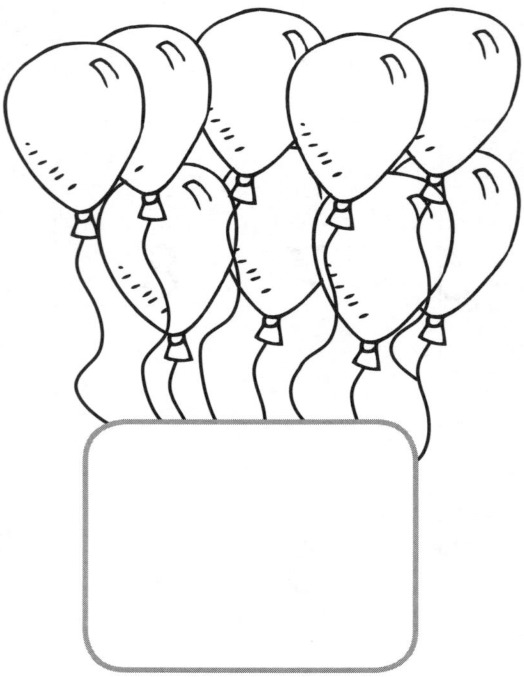

Date: ____________

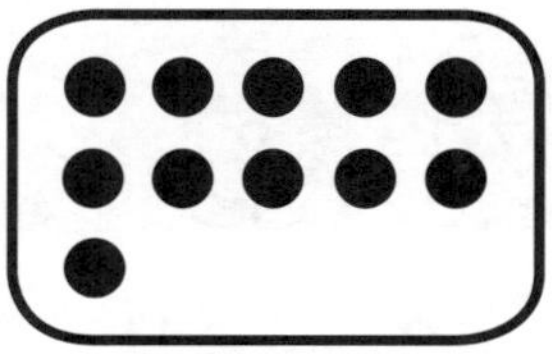

11 eleven

Write the number 11.

Circle each 11.

11 3 4

6 5

2 11

8 7

11 9

Count up!

8 9 10 ____ 12

Count back!

12 ____ 10 9 8

Fill in the ○ to show each answer.

1. How many ▲?

▲ ▲ ▲ ▲ ▲ ▲ ▲ ▲ ▲ ▲
▲

○ 10

○ 7

○ 11

2. Which numbers are missing?

○ 3 and 4

○ 5 and 6

○ 9 and 10

3. Which number comes just before 11?

○ 7

○ 10

○ 3

4. Which number is the largest?

○ 2

○ 9

○ 3

Draw 11 of something.

To parents Take a box of building blocks. Ask your child to count **11** blocks and line them up.

Date: ____________

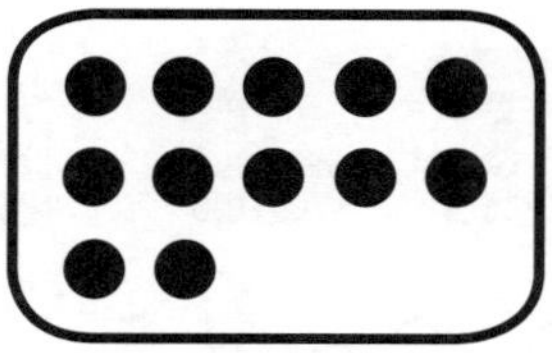

12 twelve

Write the number 12.

12 12

Circle each 12.

4

12 3

10 5

11 12

8

7

9

12

Count up!

8 9 10 11 ____

Count back!

15 14 13 ____ 11

Fill in the ○ to show each answer.

1. How many ▲?

▲ ▲ ▲ ▲ ▲ ▲ ▲ ▲ ▲ ▲
▲ ▲

○ 12

○ 10

○ 8

2. Which numbers are missing?

○ 2 and 5

○ 9 and 12

○ 3 and 7

3. Which number comes just before 12?

○ 11

○ 9

○ 3

4. Which number is the largest?

○ 7

○ 5

○ 11

Draw 12 of something.

To parents Give your child an empty egg carton and some small items. Ask her to put one item in each of the **12** carton cups and count the total number of items.

Date: ____________

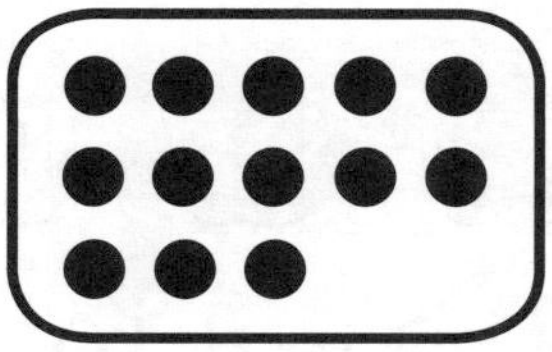

13 thirteen

Write the number 13.

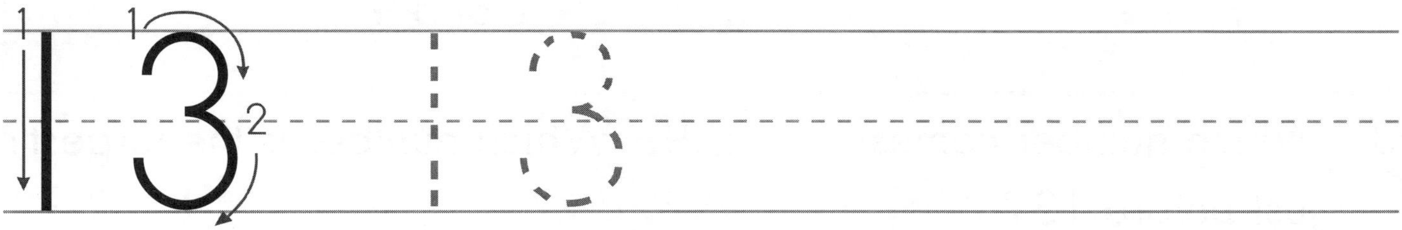

Circle each 13.

13 3 4

10 5

12 13

8 7

13 9

Count up!

9 10 11 12 ___

Count back!

15 14 ___ 12 11

Fill in the ○ to show each answer.

1. How many ▲?

▲ ▲ ▲ ▲ ▲ ▲ ▲ ▲ ▲ ▲
▲ ▲ ▲

○ 12

○ 13

○ 7

2. Which numbers are missing?

○ 9 and 10

○ 5 and 6

○ 2 and 4

3. Which number comes just before 13?

○ 9

○ 10

○ 12

4. Which number is the largest?

○ 11

○ 8

○ 2

Draw 13 of something.

To parents Give your child some play dough and ask her to make **13** small balls. Put the balls together in a line to resemble a caterpillar.

Date: ____________

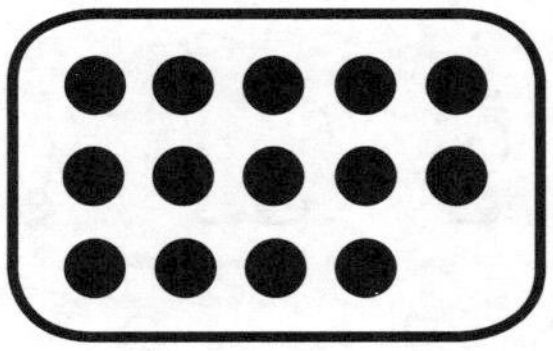 14 fourteen

Write the number 14.

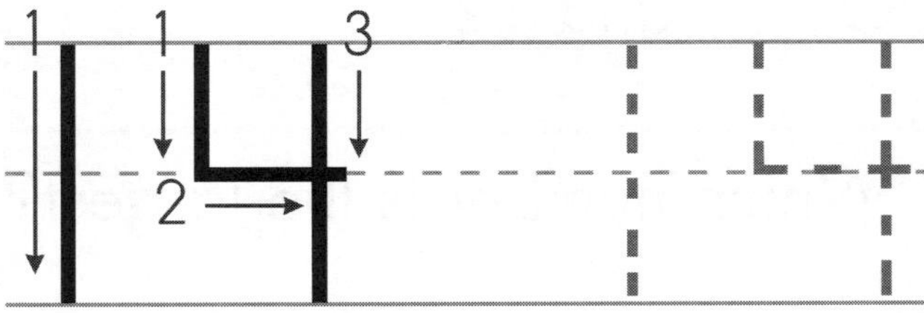

Circle each 14.

14 3 4

11 5

12 14

8 7

14 9

Count up!

11 12 13 ____ 15

Count back!

17 16 15 ____ 13

Fill in the ○ to show each answer.

1. How many ▲?

▲ ▲ ▲ ▲ ▲ ▲ ▲ ▲ ▲ ▲
▲ ▲ ▲ ▲

○ 10

○ 14

○ 9

2. Which numbers are missing?

○ 8 and 9

○ 2 and 3

○ 11 and 12

3. Which number comes just before 14?

○ 11

○ 13

○ 10

4. Which number is the largest?

○ 12

○ 6

○ 11

Draw 14 of something.

To parents Go on a scavenger hunt around the house with your child and look for **14** items of the same color.

Date: ______________

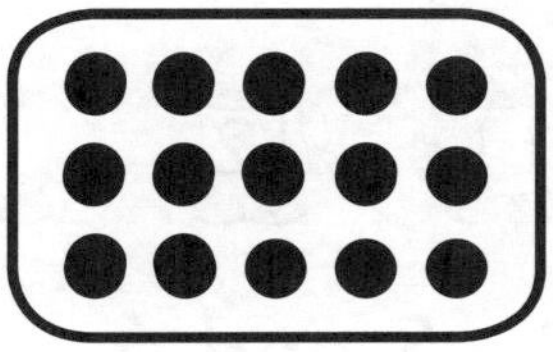

15 fifteen

Write the number 15.

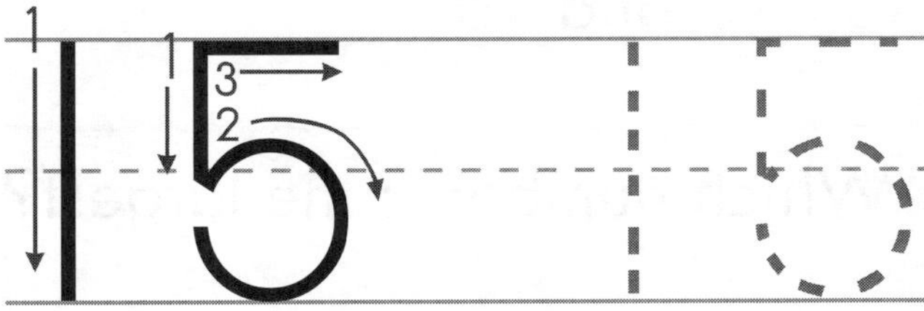

Circle each 15.

15 3 4

11 5

15 12

8 7

15 9

Count up!

13 14 ____ 16 17

Count back!

____ 14 13 12 11

Fill in the ○ to show each answer.

1. How many ▲?

▲ ▲ ▲ ▲ ▲ ▲ ▲ ▲ ▲ ▲
▲ ▲ ▲ ▲ ▲

○ 10

○ 15

○ 5

2. Which numbers are missing?

○ 5 and 6

○ 8 and 9

○ 13 and 15

3. Which number comes just before 15?

○ 14

○ 10

○ 7

4. Which number is the largest?

○ 7

○ 13

○ 10

Draw 15 of something.

To parents	Give your child a box of raisins. Ask her to count and pick out **15** raisins. Have her count the raisins again as she eats them.

Date: ____________

Review Numbers 11 to 15

Circle the number that tells how many.

12 14 6	
12 7 8	
9 5 11	
15 5 11	
3 10 13	

Count and write the number in the box.
Color the set that has **more** balls.

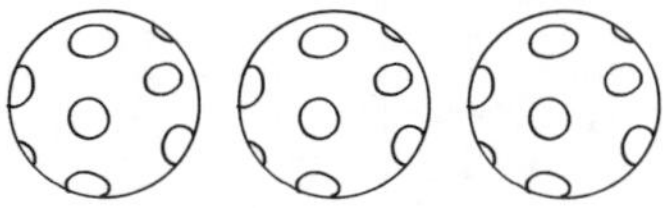

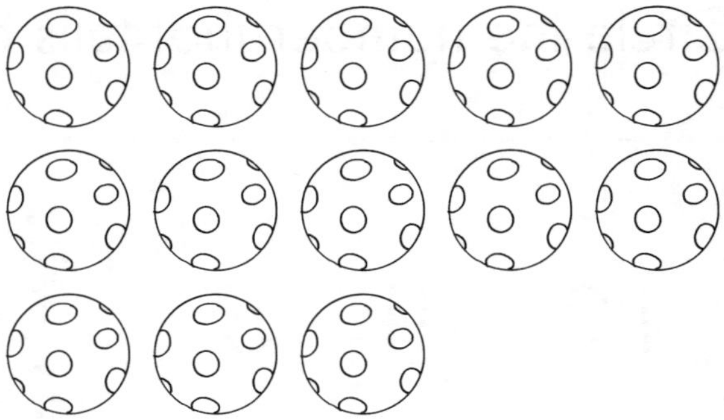

Count and write the number in the box.
Color the set that has **fewer** balls.

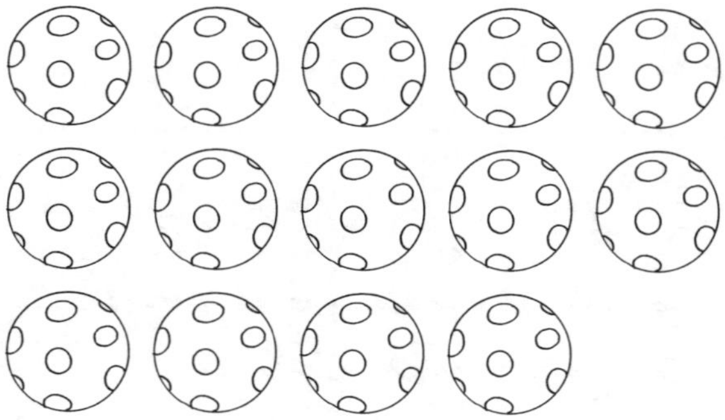

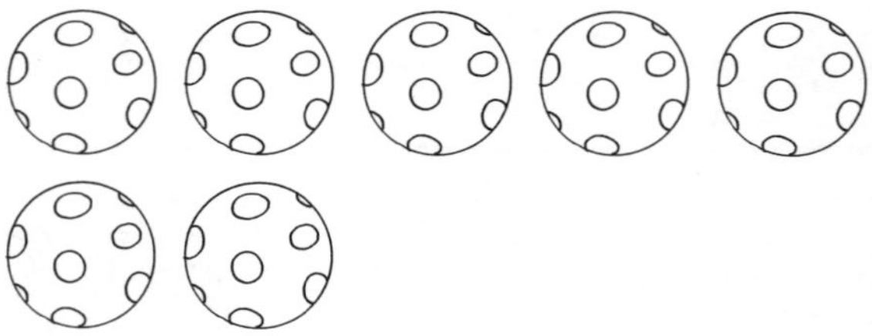

Date: ____________

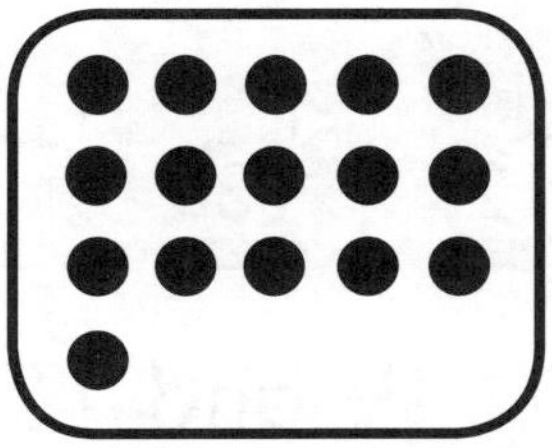 16 sixteen

Write the number 16.

16 16

Circle each 16.

16 3 4

16 5

16 11

8 7

15 9

Count up!

12 13 14 15 ____

Count back!

____ 15 14 13 12

Fill in the ○ to show each answer.

1. How many ▲?

▲ ▲ ▲ ▲ ▲ ▲ ▲ ▲ ▲ ▲
▲ ▲ ▲ ▲ ▲ ▲

○ 15

○ 2

○ 16

2. Which numbers are missing?

○ 14 and 15

○ 9 and 10

○ 7 and 8

3. Which number comes just before 16?

○ 11

○ 15

○ 3

4. Which number is the largest?

○ 14

○ 4

○ 11

Draw 16 of something.

To parents Give your child a long piece of yarn, knotted at one end, and a box of small items such as beads or paper clips. Ask her to count and pick out **16** small items and string them on the yarn.

Date: ____________

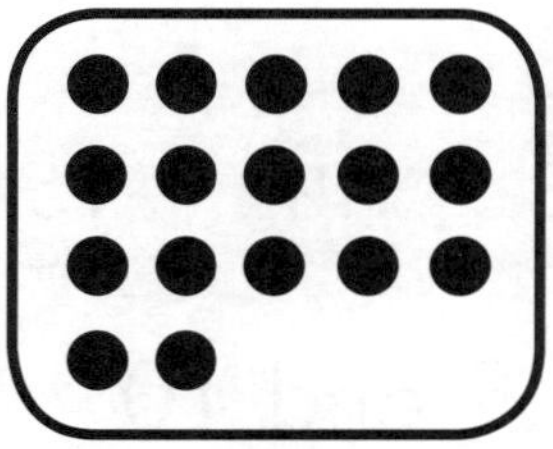

17 seventeen

Write the number 17.

17

Circle each 17.

17 3 4

13 5

17 12

8 7

17 9

Count up!

14 15 16 ____ 18

Count back!

20 19 18 ____ 16

Fill in the ○ to show each answer.

1. How many ▲?

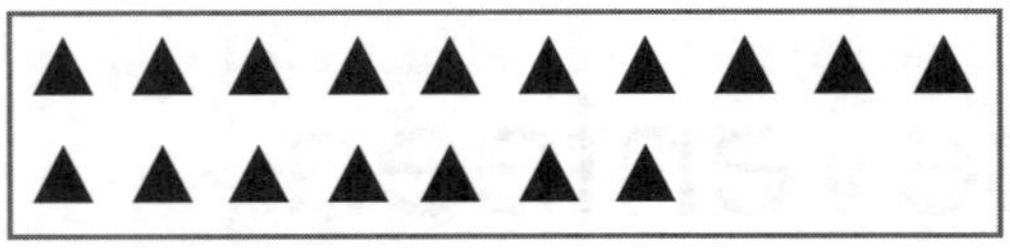

○ 17

○ 14

○ 8

2. Which numbers are missing?

○ 9 and 10

○ 16 and 17

○ 4 and 11

3. Which number comes just before 17?

○ 4

○ 12

○ 16

4. Which number is the largest?

○ 15

○ 9

○ 14

Draw 17 of something.

To parents Give your child a pack of toothpicks. Ask her to count and take out **17** toothpicks, and place them in a row on a table.

Date: ______________

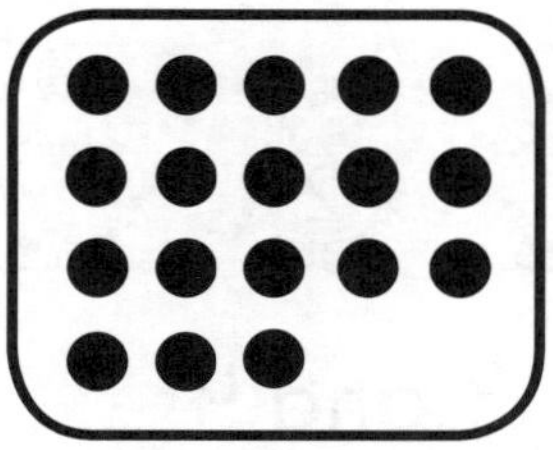

18 eighteen

Write the number 18.

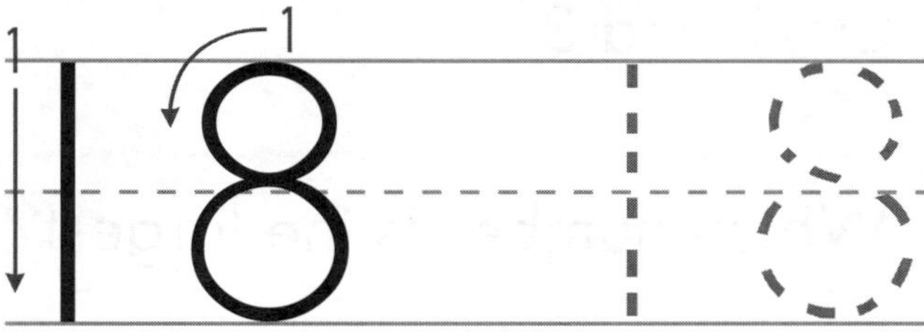

Circle each 18.

18 3 4

18

13

5 12

18 7

9

8

Count up!

16 17 ____ 19 20

Count back!

19 ____ 17 16 15

Fill in the ○ to show each answer.

1. How many ▲?

▲ ▲ ▲ ▲ ▲ ▲ ▲ ▲ ▲ ▲
▲ ▲ ▲ ▲ ▲ ▲ ▲ ▲

○ 7

○ 18

○ 10

2. Which numbers are missing?

○ 3 and 4

○ 9 and 10

○ 1 and 3

3. Which number comes just before 18?

○ 3

○ 17

○ 9

4. Which number is the largest?

○ 18

○ 10

○ 1

Draw 18 of something.

To parents	Have your child count and pick out **18** paper clips and link these together to make a chain. Have her count the paper clips again as she links them.

Date: ______________

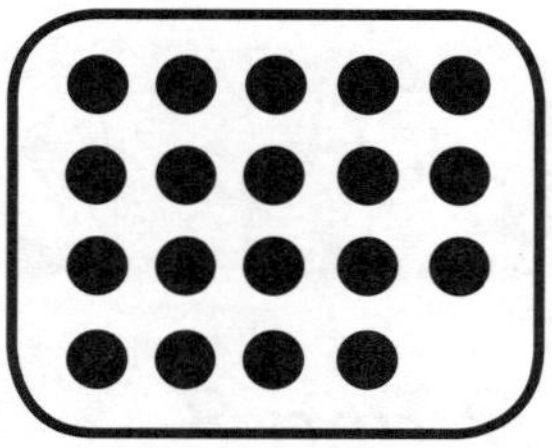

19 nineteen

Write the number 19.

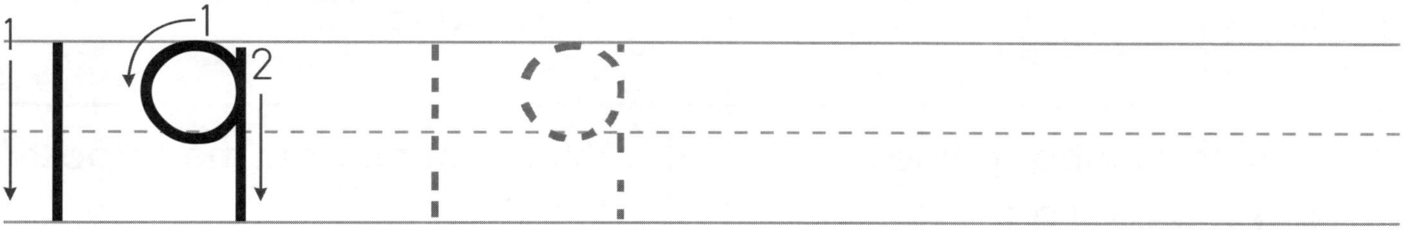

Circle each 19.

18 3 4

18

19

5 19

19 7

9

8

Count up!

16 17 18 ____ 20

Count back!

20 ____ 18 17 16

Fill in the ○ to show each answer.

1. How many ▲?

▲ ▲ ▲ ▲ ▲ ▲ ▲ ▲ ▲ ▲
▲ ▲ ▲ ▲ ▲ ▲ ▲ ▲ ▲

○ 19

○ 2

○ 11

2. Which numbers are missing?

○ 2 and 8

○ 16 and 18

○ 12 and 13

3. Which number comes just before 19?

○ 18

○ 14

○ 16

4. Which number is the largest?

○ 7

○ 5

○ 15

Draw 19 of something.

To parents Make a small photo frame using ice-cream sticks. Let your child paste **19** pieces of dry pasta all around the frame using craft glue.

Date: ____________

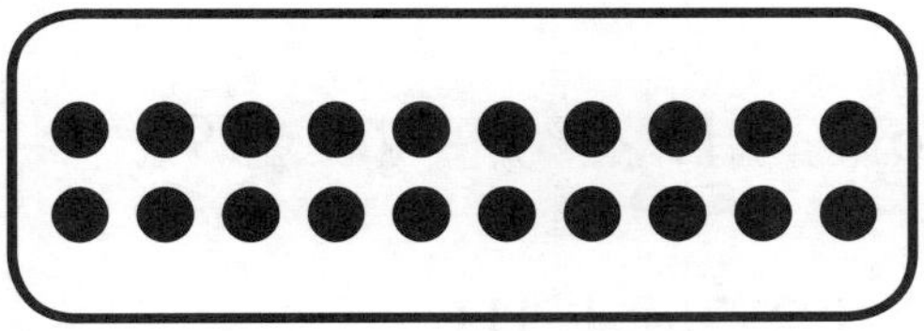

20 twenty

Write the number 20.

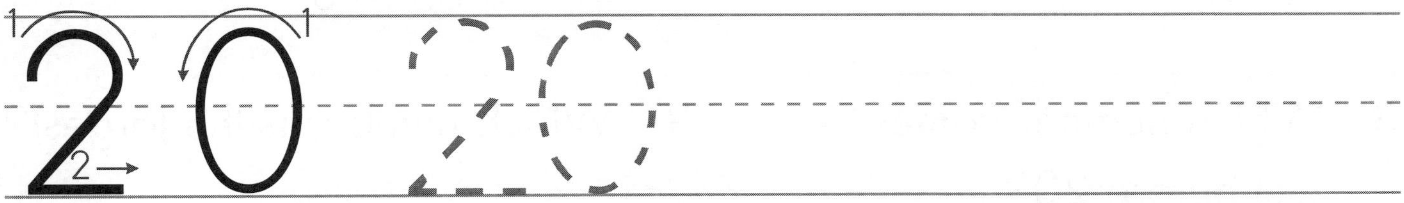

Circle each 20.

18 3 4

10

20

5 16

13

7

20

20

Count up!

16 17 18 19 ____

Count back!

____ 19 18 17 16

Fill in the ○ to show each answer.

1. How many ▲?

▲▲▲▲▲▲▲▲▲▲
▲▲▲▲▲▲▲▲▲▲

○ 15

○ 10

○ 20

2. Which numbers are missing?

○ 13 and 14

○ 8 and 9

○ 18 and 20

3. Which number comes just before 20?

○ 19

○ 17

○ 11

4. Which number is the largest?

○ 14

○ 18

○ 9

Draw 20 of something.

To parents Ask your child to count her fingers and toes. Then point out that 10 fingers and 10 toes add up to **20**.

Date: ____________

Review Numbers 16 to 20

Circle the number that tells how many.

2 12 19	
16 9 5	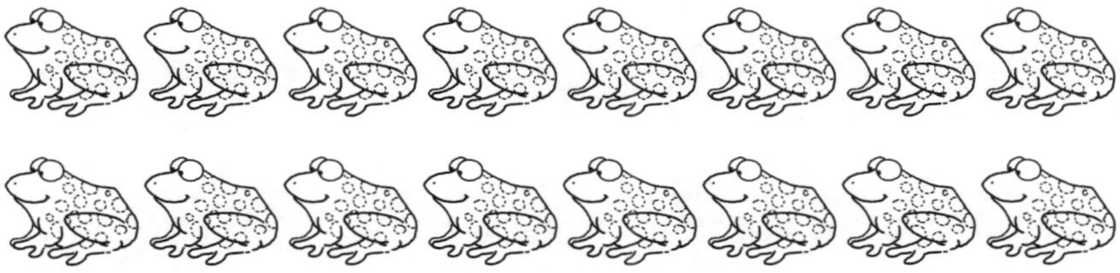
8 18 12	
9 17 13	
8 20 17	

Count and write the number in the box.
Color the set that has **more** balls.

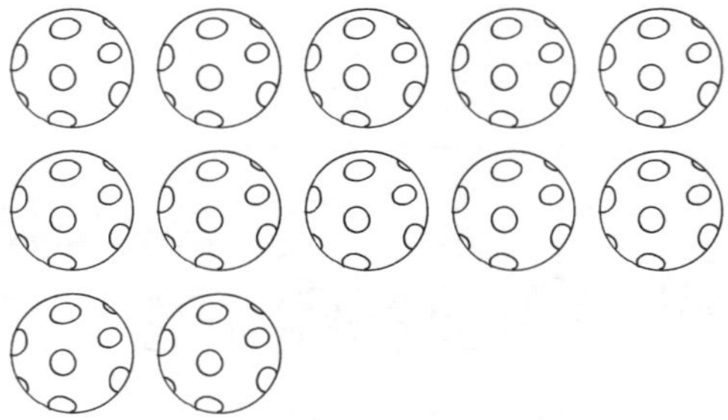

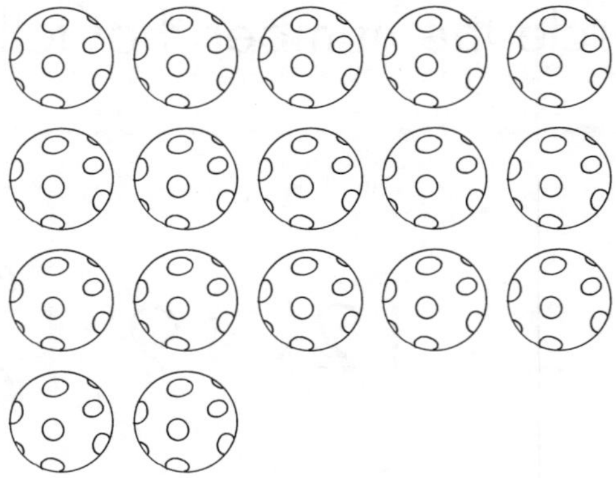

Count and write the number in the box.
Color the set that has **fewer** balls.

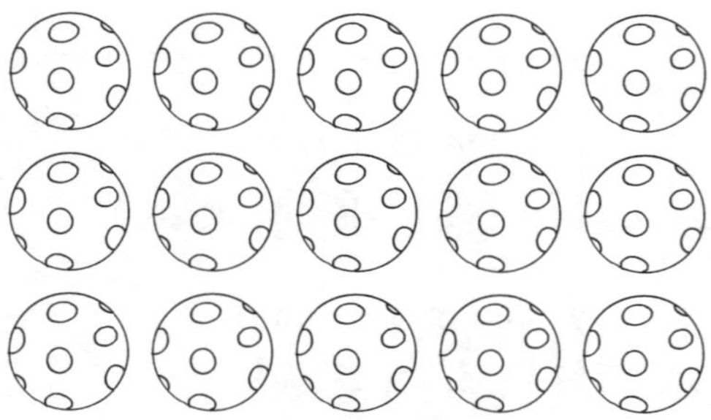

Date: ______________

21 twenty-one

Write the number 21.

21 2

Circle each 21.

18 3 4

21

21

5 19

21 7

9

8

Count up!

18 19 20 ____ 22

Count back!

22 ____ 20 19 18

Fill in the ○ to show each answer.

1. How many ▲?

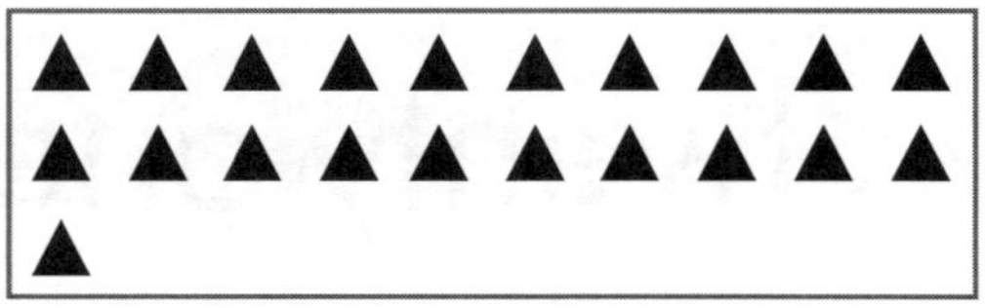

○ 20

○ 5

○ 21

2. Which numbers are missing?

○ 9 and 11

○ 13 and 16

○ 19 and 20

3. Which number comes just before 21?

○ 20

○ 9

○ 17

4. Which number is the largest?

○ 14

○ 19

○ 4

Draw 21 of something.

To parents Give your child a box of cherries. Ask her to count and pick out **21** cherries, and place them in a line on a clean table. Have her count the cherries again as she eats them.

Date: ____________

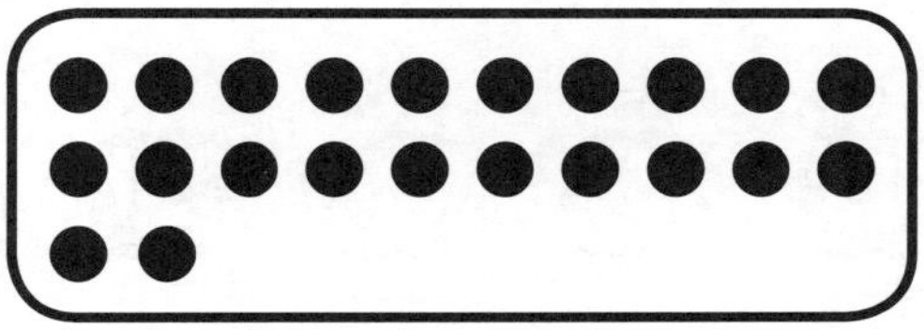

22 twenty-two

Write the number 22.

22 22

Circle each 22.

4

10 3

18

22

5 22

13

7

22

8

Count up!

20 21 ____ 23 24

Count back!

24 23 ____ 21 20

Fill in the ○ to show each answer.

1. How many ▲?

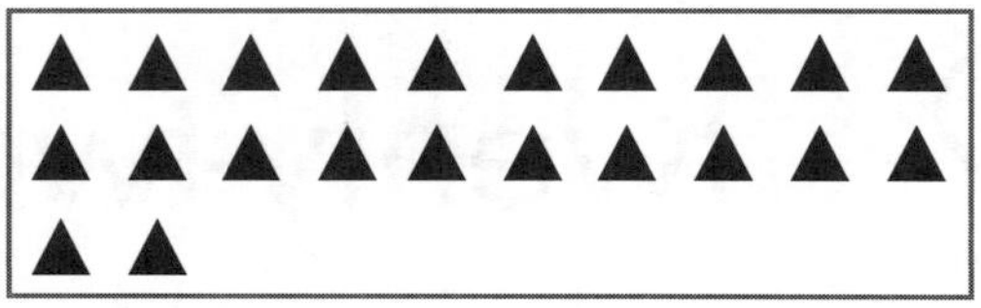

○ 22

○ 5

○ 12

2. Which numbers are missing?

○ 9 and 10

○ 19 and 20

○ 15 and 17

3. Which number comes just before 22?

○ 21

○ 8

○ 18

4. Which number is the largest?

○ 19

○ 10

○ 20

Draw 22 of something.

To parents Give your child **22** beads and ask her to string them into a necklace. Have her count each bead as she strings it.

Date: ________________

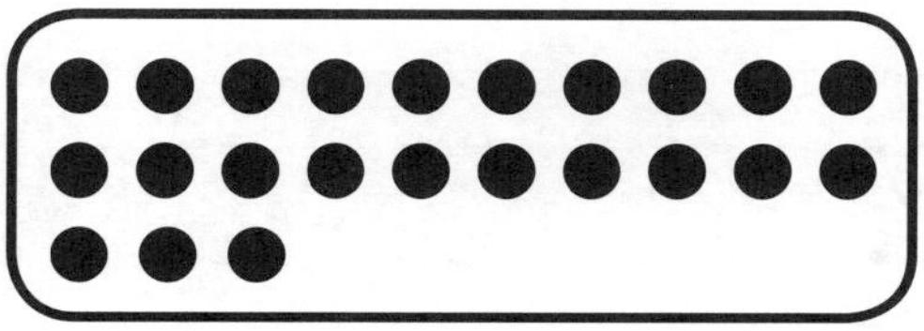

23 twenty-three

Write the number 23.

Circle each 23.

10 3 4

18

23

5 23

23 7

20

8

Count up!

19 20 21 22 ____

Count back!

24 ____ 22 21 20

Fill in the ○ to show each answer.

1. How many ▲?

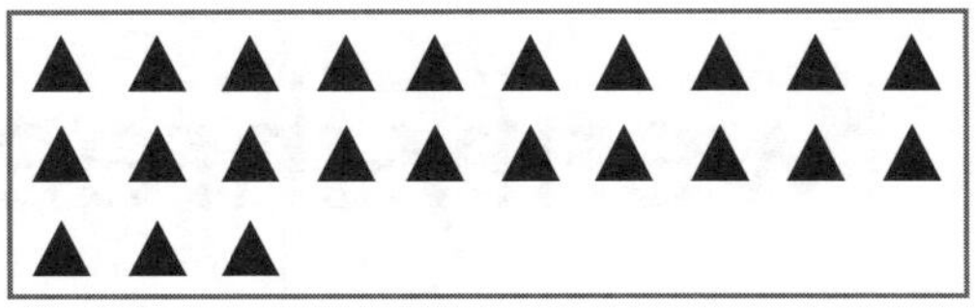

○ 20

○ 23

○ 7

2. Which numbers are missing?

○ 21 and 22

○ 11 and 13

○ 6 and 7

3. Which number comes just before 23?

○ 6

○ 17

○ 22

4. Which number is the largest?

○ 21

○ 1

○ 13

Draw 23 of something.

To parents Have your child count and pick out **23** paper clips and link these together to make a paper clip chain. Have her count the paper clips again as she links them.

Date: ____________

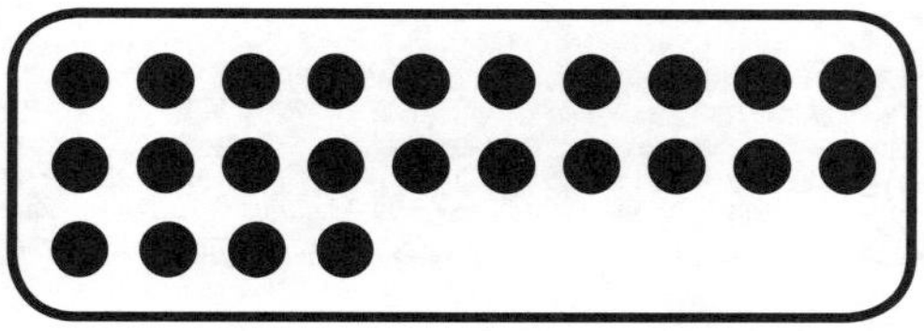

24 twenty-four

Write the number 24.

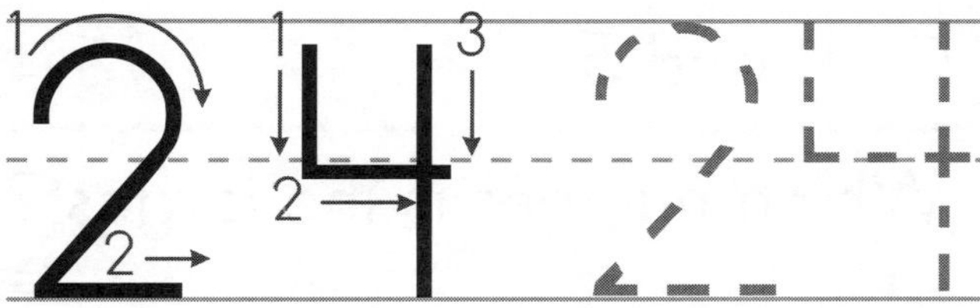

Circle each 24.

10 3 4

18

24

5 24

24

7

20

8

Count up!

21 22 23 ____ 25

Count back!

26 25 ____ 23 22

Fill in the ○ to show each answer.

1. How many ▲?

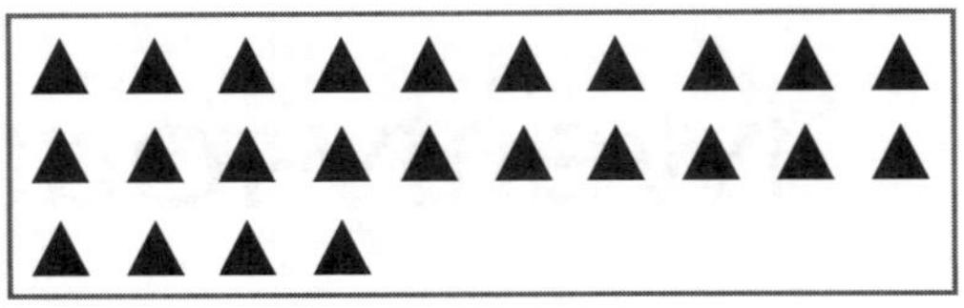

○ 24

○ 21

○ 11

2. Which numbers are missing?

○ 10 and 12

○ 21 and 23

○ 4 and 5

3. Which number comes just before 24?

○ 9

○ 20

○ 23

4. Which number is the largest?

○ 20

○ 22

○ 8

Draw 24 of something.

To parents Give your child a long piece of yarn and show her how to make a simple knot. Ask her to knot the piece of yarn **24** times.

Date: ____________

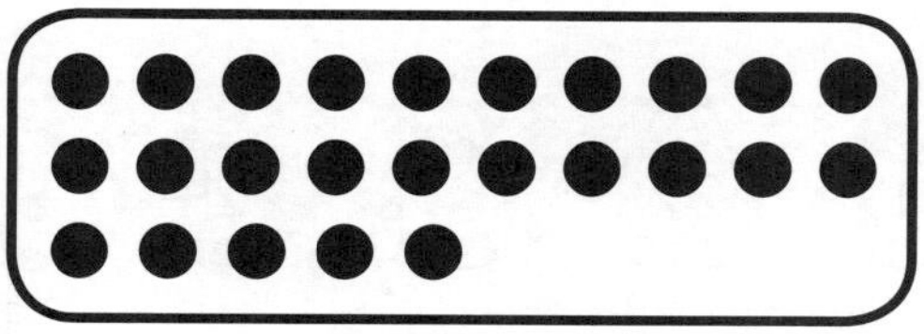

25 twenty-five

Write the number 25.

25 25

Circle each 25.

10 3 4

25

25

5 20

25 7

8 20

Count up!

23 24 ____ 26 27

Count back!

____ 24 23 22 21

Fill in the ○ to show each answer.

1. How many ▲?

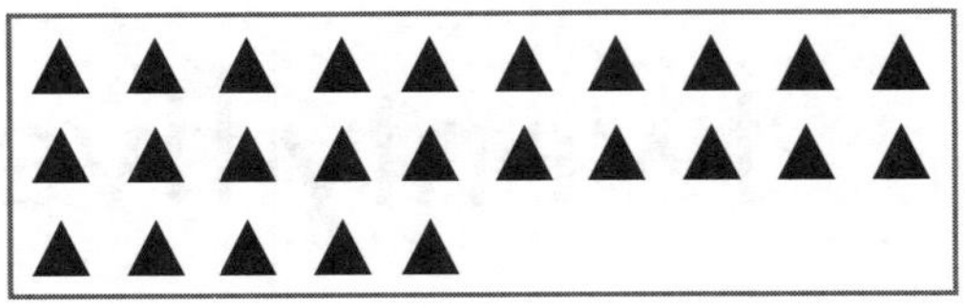

○ 20

○ 11

○ 25

2. Which numbers are missing?

○ 23 and 25

○ 10 and 12

○ 19 and 20

3. Which number comes just before 25?

○ 16

○ 24

○ 21

4. Which number is the largest?

○ 23

○ 7

○ 12

Draw 25 of something.

To parents Take a small wooden bowl or tray. Have your child decorate it by pasting **25** pieces of dried pasta on it.

Date: ____________

Review Numbers 21 to 25

Circle the number that tells how many.

24 20 18	
11 23 17	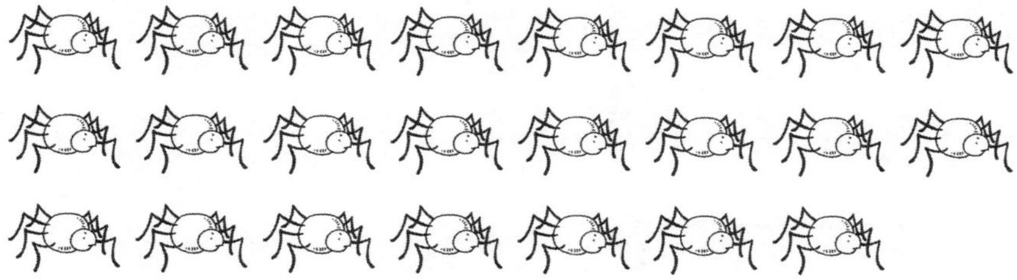
13 25 19	
21 10 19	
18 22 12	

Count and write the number in the box.
Color the set that has **more** crayons.

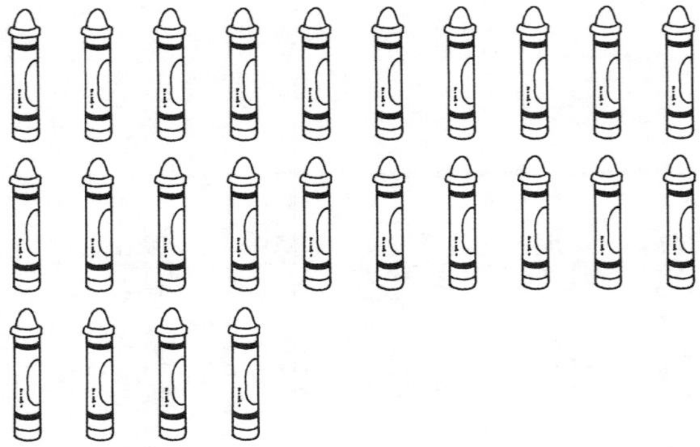

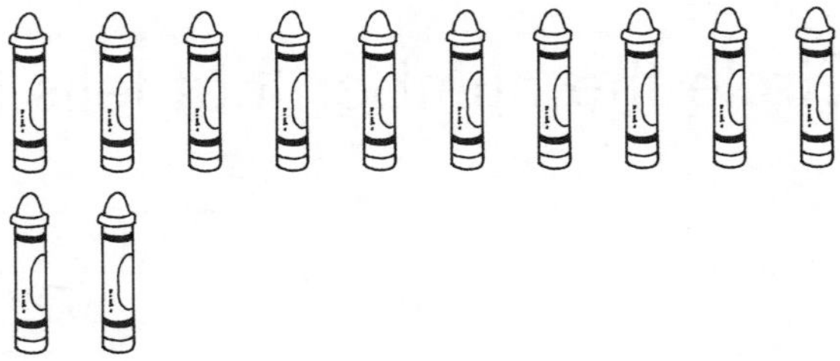

Count and write the number in the box.
Color the set that has **fewer** crayons.

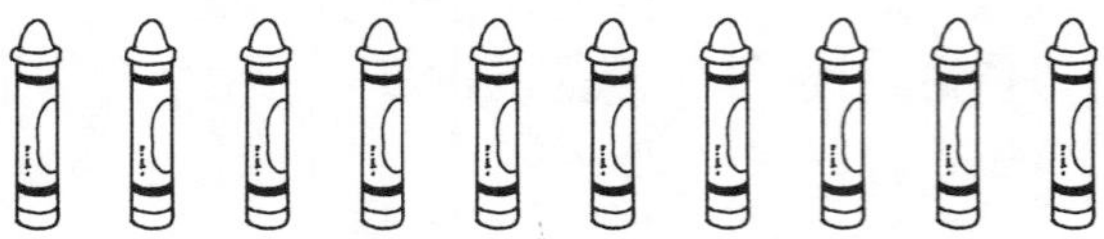

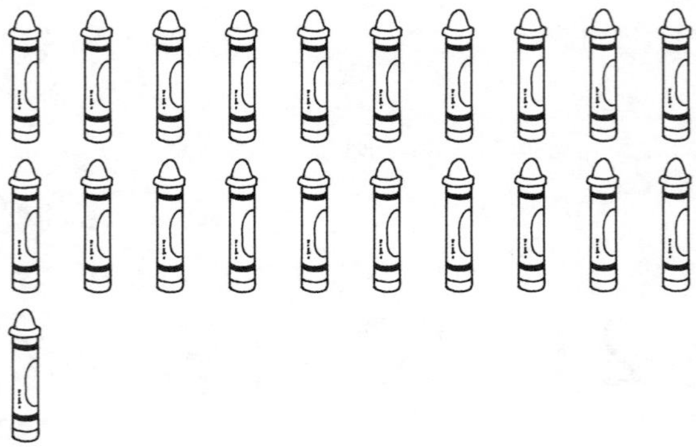

Date: ______________

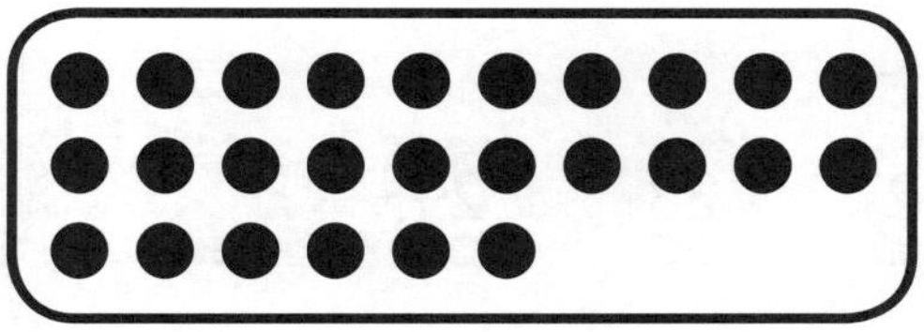

26 twenty-six

Write the number 26.

26 26

Circle each 26.

4

10 3

18

21

5 26

26 7

26

8

Count up!

22 23 24 25 ____

Count back!

27 ____ 25 24 23

Fill in the ○ to show each answer.

1. How many ▲?

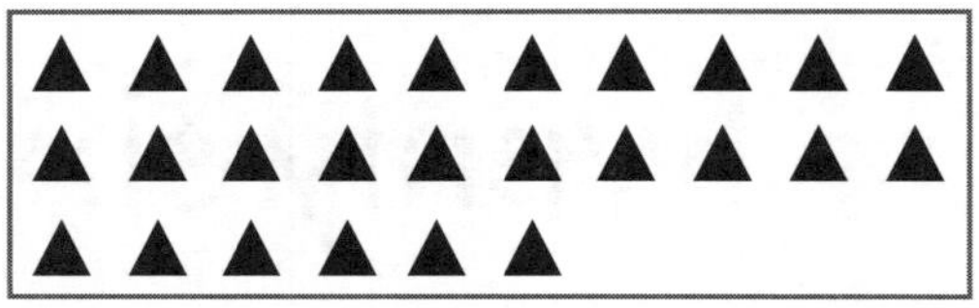

○ 15

○ 26

○ 6

2. Which numbers are missing?

○ 19 and 20

○ 12 and 13

○ 25 and 26

3. Which number comes just before 26?

○ 25

○ 21

○ 12

4. Which number is the largest?

○ 18

○ 7

○ 24

Draw 26 of something.

To parents	Give your child a pack of rubber bands. Ask her to count and take out **26** rubber bands, and place them in a row on a table.

Date: ____________

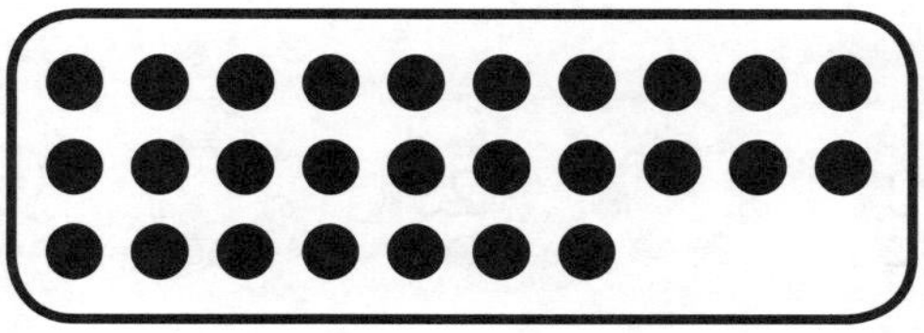

27 twenty-seven

Write the number 27.

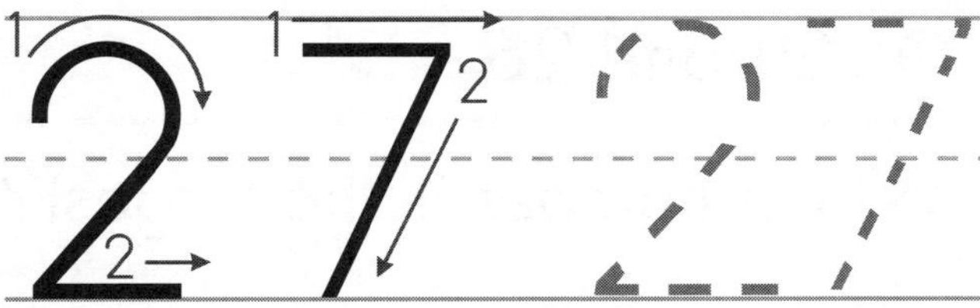

Circle each 27.

10 3 4

18

27

5 27

27 7

23

8

Count up!

25 26 ____ 28 29

Count back!

30 29 28 ____ 26

Fill in the ○ to show each answer.

1. How many ▲?

▲▲▲▲▲▲▲▲▲▲
▲▲▲▲▲▲▲▲▲▲
▲▲▲▲▲▲▲

○ 10

○ 20

○ 27

2. Which numbers are missing?

○ 12 and 14

○ 8 and 9

○ 24 and 26

3. Which number comes just before 27?

○ 20

○ 26

○ 19

4. Which number is the largest?

○ 25

○ 5

○ 21

Draw 27 of something.

To parents Go on a scavenger hunt around the house with your child and look for **27** items of the same color.

Date: ____________

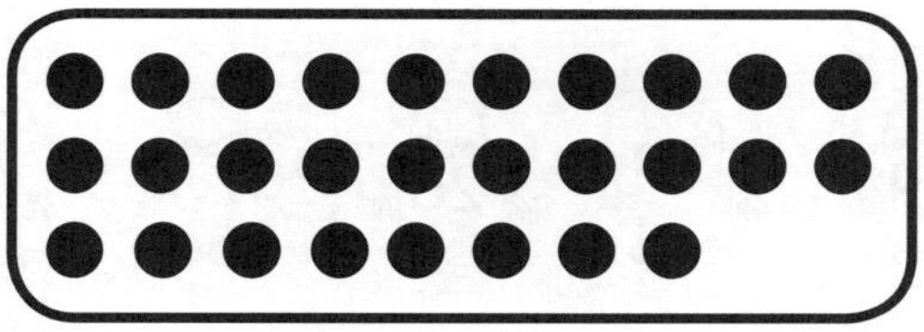

28 twenty-eight

Write the number 28.

28

Circle each 28.

10 3 4

16 18

5 28

28 7

8 28

Count up!

25 26 27 ____ 29

Count back!

30 29 ____ 27 26

Fill in the ○ to show each answer.

1. How many ▲?

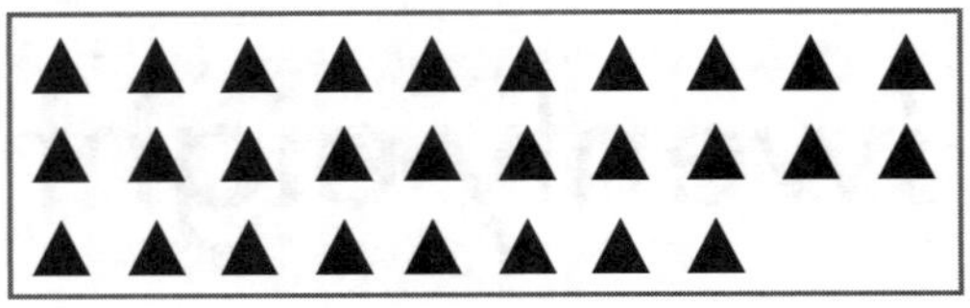

○ 15

○ 28

○ 4

2. Which numbers are missing?

○ 25 and 27

○ 14 and 15

○ 5 and 6

3. Which number comes just before 28?

○ 16

○ 25

○ 27

4. Which number is the largest?

○ 26

○ 16

○ 3

Draw 28 of something.

To parents Stand at a safe place that overlooks a busy street. Have your child count the number of cars that go by. Stop when she reaches **28**.

Date: ______________

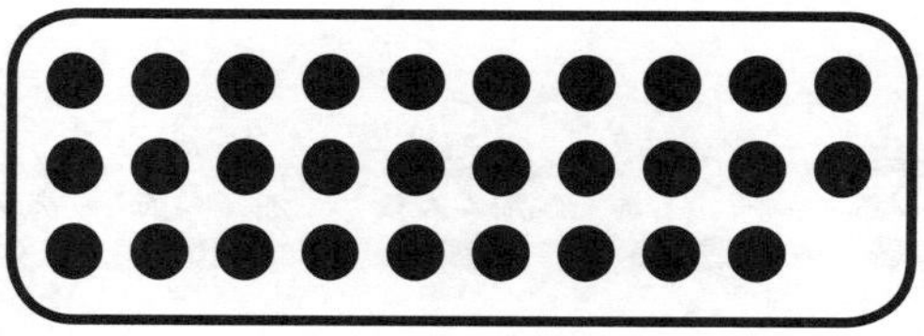

29 twenty-nine

Write the number 29.

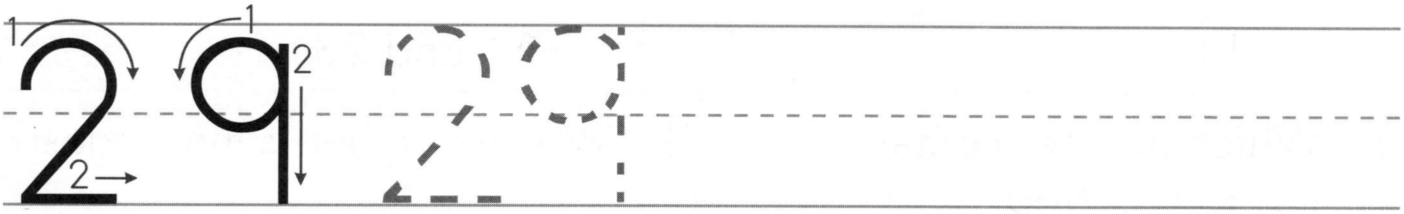

Circle each 29.

10 3 4

18

29

5 29

20

7

29

8

Count up!

26 27 28 ____ 30

Count back!

30 ____ 28 27 26

Fill in the ○ to show each answer.

1. How many ▲?

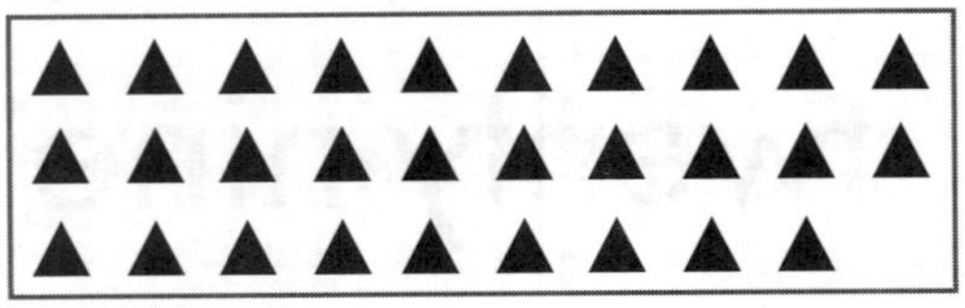

○ 29

○ 4

○ 11

2. Which numbers are missing?

○ 6 and 8

○ 26 and 28

○ 21 and 23

3. Which number comes just before 29?

○ 13

○ 28

○ 24

4. Which number is the largest?

○ 9

○ 12

○ 27

Draw 29 of something.

To parents Give your child **29** building blocks and ask her to line them up. Have her count each block as she arranges them.

Date: ______________

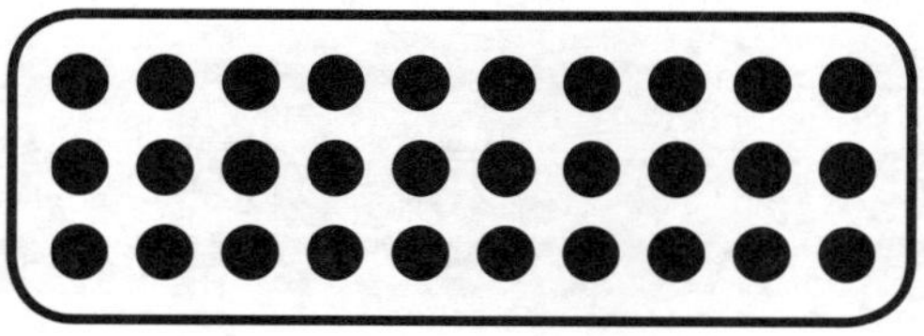

30 thirty

Write the number 30.

30 30

Circle each 30.

10 3 4

18

30

5 30

30 7

20

8

Count up!

26 27 28 29 ____

Count back!

____ 29 28 27 26

Fill in the ○ to show each answer.

1. How many ▲?

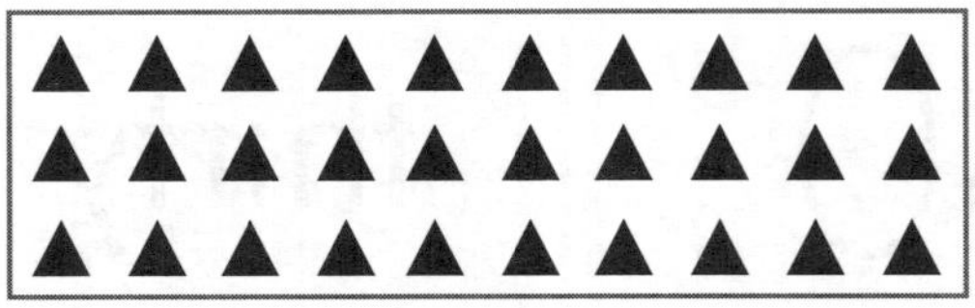

○ 20

○ 10

○ 30

2. Which numbers are missing?

○ 27 and 30

○ 15 and 18

○ 8 and 9

3. Which number comes just before 30?

○ 25

○ 29

○ 11

4. Which number is the largest?

○ 6

○ 28

○ 14

Draw 30 of something.

To parents Play a game of Hide-and-Seek with your child. Have her count up to **30** while you look for a place to hide. Ask her to look for you when she is done counting.

Review Numbers 26 to 30

Circle the number that tells how many.

26 16 20	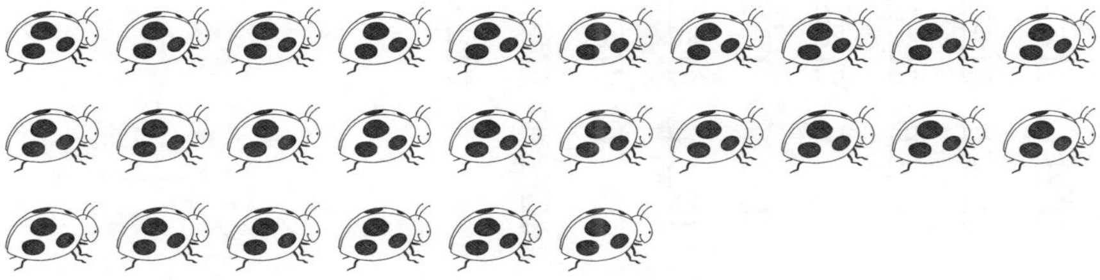
10 20 30	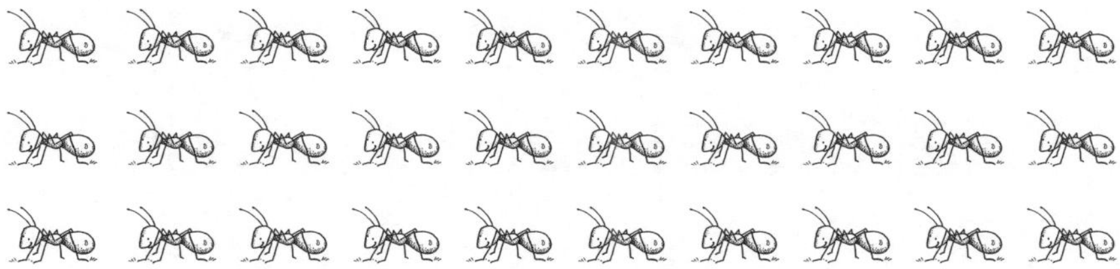
7 27 21	
29 2 19	
8 28 12	

Count and write the number in the box.
Color the set that has **more** crayons.

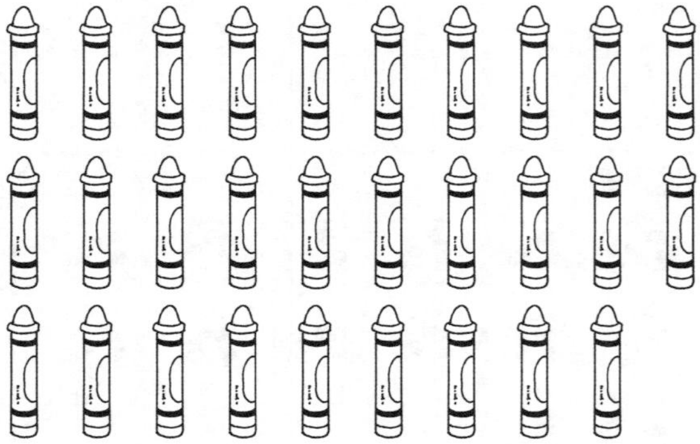

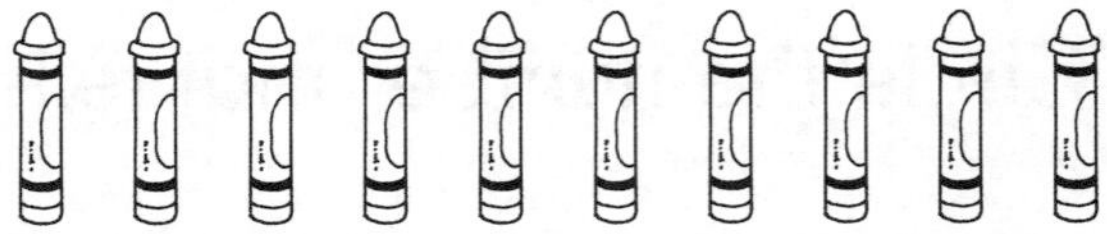

Count and write the number in the box.
Color the set that has **fewer** crayons.

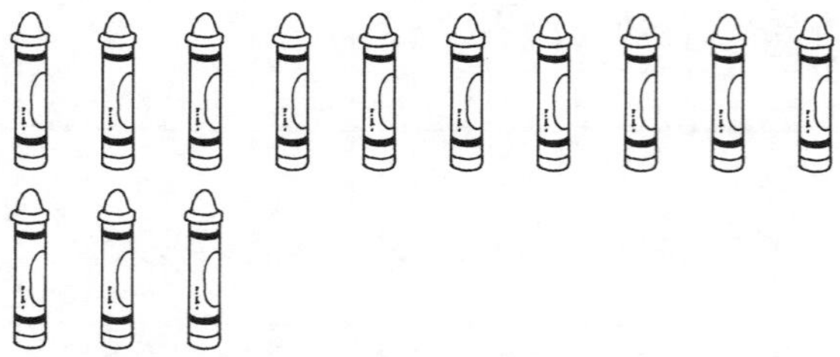

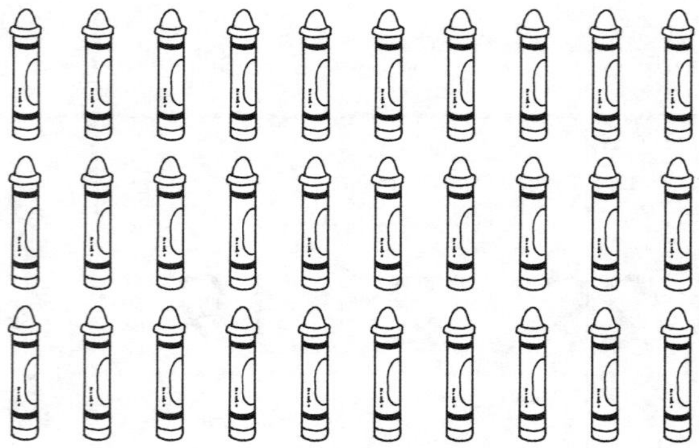

Answer Key

Page 6

1. Which set has 1?
   ● (butterfly)
2. Which set has more than 1?
   ● (3 muffins)
3. Which number comes just before 2?
   ● 1
4. Which number is missing?
   ● 1

Page 8

1. Which set has 2?
   ● (2 strawberries)
2. Which set has more than 2?
   ● (3 dogs)
3. Which number comes just before 3?
   ● 2
4. Which numbers are missing?
   ● 1 and 2

Page 10

1. Which set has 3?
   ● (3 ducks)
2. Which set has more than 2?
   ● (3 airplanes)
3. Which number comes just before 3?
   ● 2
4. Which numbers are missing?
   ● 2 and 3

Page 12

1. Which set has 4?
   ● (4 cows)
2. Which set has more than 3?
   ● (4 drums)
3. Which number comes just before 4?
   ● 3
4. Which numbers are missing?
   ● 2 and 4

Page 14

1. Which set has 5?
   ● (5 bears)
2. Which set has more than 3?
   ● (4 bees)
3. Which number comes just before 5?
   ● 4
4. Which numbers are missing?
   ● 3 and 5

Page 15

3; 1; 5; 4; 2

Page 16

Count and write the number in the box.
Color the set that has **more** balloons.

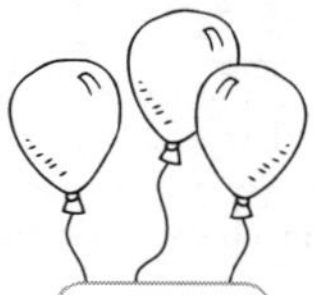

3

Color this set.

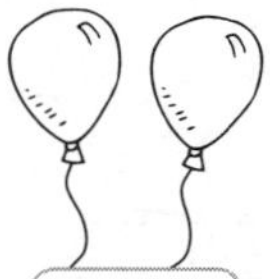

2

Count and write the number in the box.
Color the set that has **fewer** balloons.

3

Color this set.

5

Page 18

1. Which set has 6?
   ● (6 kangaroos)
2. Which set has more than 4?
   ● (5 chicks)
3. Which number comes just before 6?
   ● 5
4. Which numbers are missing?
   ● 4 and 5

Page 20

1. Which set has 7?
   ● (7 apples)
2. Which set has more than 5?
   ● (6 pies)
3. Which number comes just before 7?
   ● 6
4. Which numbers are missing?
   ● 4 and 5

Page 22

1. Which set has 8?
   ● (8 fish)
2. Which set has more than 6?
   ● (7 turtles)
3. Which number comes just before 8?
   ● 7
4. Which numbers are missing?
   ● 6 and 8

Page 24

1. Which set has 9?
   - 
2. Which set has more than 7?
   - 
3. Which number comes just before 9?
   - 8
4. Which numbers are missing?
   - 5 and 6

Page 26

1. How many ▲?
   - 10
2. Which numbers are missing?
   - 1 and 2
3. Which number comes just before 10?
   - 9
4. Which number is the largest?
   - 10

Page 27

6; 10; 9; 8; 7

Page 28

Count and write the number in the box.
Color the set that has **more** balloons.

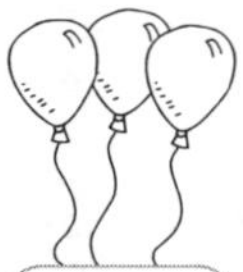

3

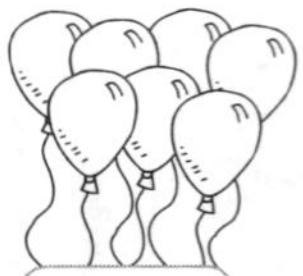

7

Color this set.

Count and write the number in the box.
Color the set that has **fewer** balloons.

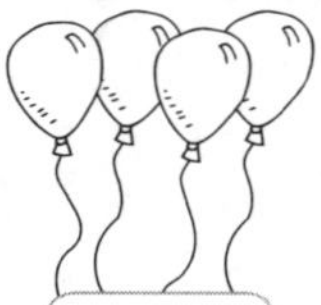

4

Color this set.

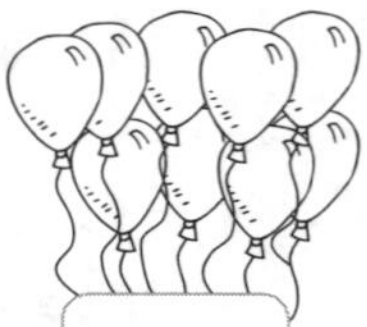

9

Page 30

1. How many ▲?
   - 11
2. Which numbers are missing?
   - 9 and 10
3. Which number comes just before 11?
   - 10
4. Which number is the largest?
   - 9

Page 32

1. How many ▲?
   - 12
2. Which numbers are missing?
   - 9 and 12
3. Which number comes just before 12?
   - 11
4. Which number is the largest?
   - 11

Page 34

1. How many ▲?
   - 13
2. Which numbers are missing?
   - 9 and 10
3. Which number comes just before 13?
   - 12
4. Which number is the largest?
   - 11

Page 36

1. How many ▲?
   - 14
2. Which numbers are missing?
   - 11 and 12
3. Which number comes just before 14?
   - 13
4. Which number is the largest?
   - 12

Page 38

1. How many ▲?
   - 15
2. Which numbers are missing?
   - 13 and 15
3. Which number comes just before 15?
   - 14
4. Which number is the largest?
   - 13

Page 39

14; 12; 11; 15; 13

Page 40

Count and write the number in the box.
Color the set that has **more** balls.

3

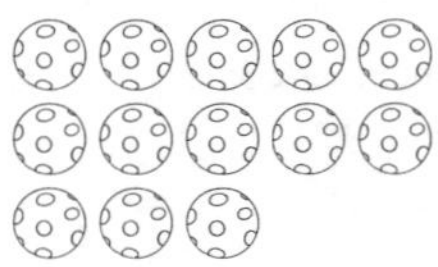

13

Color this set.

Count and write the number in the box.
Color the set that has **fewer** balls.

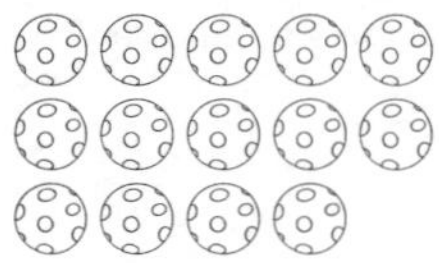

14

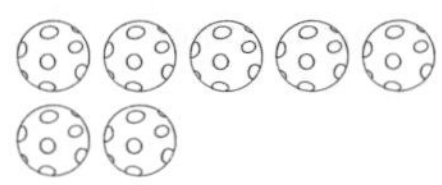

7

Color this set.

Page 42

1. How many ▲? ● 16	2. Which numbers are missing? ● 14 and 15
3. Which number comes just before 16? ● 15	4. Which number is the largest? ● 14

Page 44

1. How many ▲? ● 17	2. Which numbers are missing? ● 16 and 17
3. Which number comes just before 17? ● 16	4. Which number is the largest? ● 15

Page 46

1. How many ▲? ● 18	2. Which numbers are missing? ● 1 and 3
3. Which number comes just before 18? ● 17	4. Which number is the largest? ● 18

Page 48

1. How many ▲? ● 19	2. Which numbers are missing? ● 16 and 18
3. Which number comes just before 19? ● 18	4. Which number is the largest? ● 15

Page 50

1. How many ▲? ● 20	2. Which numbers are missing? ● 18 and 20
3. Which number comes just before 20? ● 19	4. Which number is the largest? ● 18

Page 51

19; 16; 18; 17; 20

Page 52

Count and write the number in the box.
Color the set that has **more** balls.

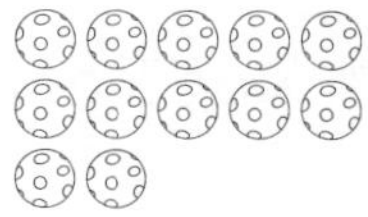

12

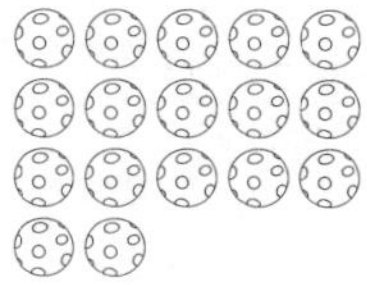

17

Color this set.

Count and write the number in the box.
Color the set that has **fewer** balls.

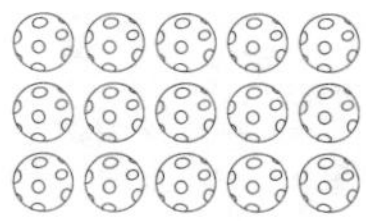

15

Color this set.

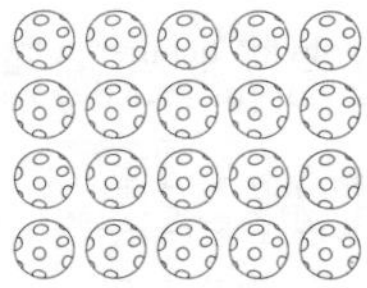

20

Page 54

1. How many ▲? ● 21	2. Which numbers are missing? ● 19 and 20
3. Which number comes just before 21? ● 20	4. Which number is the largest? ● 19

Page 56

1. How many ▲? ● 22	2. Which numbers are missing? ● 19 and 20
3. Which number comes just before 22? ● 21	4. Which number is the largest? ● 20

Page 58

1. How many ▲? ● 23	2. Which numbers are missing? ● 21 and 22
3. Which number comes just before 23? ● 22	4. Which number is the largest? ● 21

Page 60

1. How many ▲? ● 24	2. Which numbers are missing? ● 21 and 23
3. Which number comes just before 24? ● 23	4. Which number is the largest? ● 22

Page 62

1. How many ▲? ● 25	2. Which numbers are missing? ● 23 and 25
3. Which number comes just before 25? ● 24	4. Which number is the largest? ● 23

Page 63

24; 23; 25; 21; 22

Page 64

Count and write the number in the box.
Color the set that has **more** crayons.

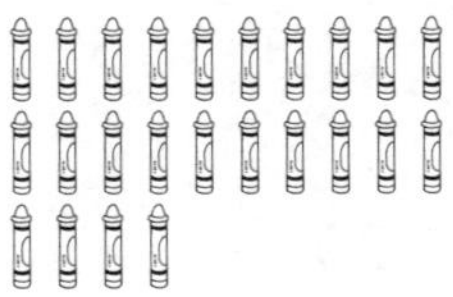

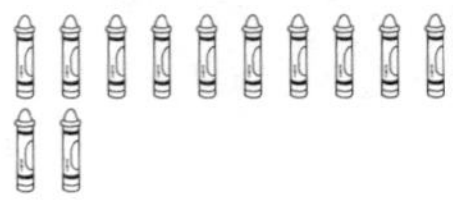

24

12

Color this set.

Count and write the number in the box.
Color the set that has **fewer** crayons.

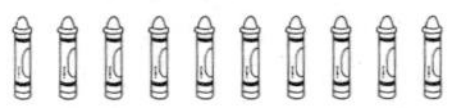

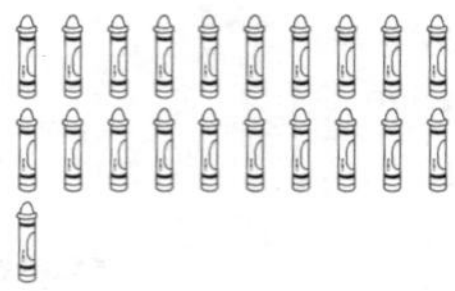

10

21

Color this set.

Page 66

1. How many ▲? ● 26	2. Which numbers are missing? ● 25 and 26
3. Which number comes just before 26? ● 25	4. Which number is the largest? ● 24

Page 68

1. How many ▲? ● 27	2. Which numbers are missing? ● 24 and 26
3. Which number comes just before 27? ● 26	4. Which number is the largest? ● 25

Page 70

1. How many ▲? ● 28	2. Which numbers are missing? ● 25 and 27
3. Which number comes just before 28? ● 27	4. Which number is the largest? ● 26

Page 72

1. How many ▲? ● 29	2. Which numbers are missing? ● 26 and 28
3. Which number comes just before 29? ● 28	4. Which number is the largest? ● 27

Page 74

1. How many ▲? ● 30	2. Which numbers are missing? ● 27 and 30
3. Which number comes just before 30? ● 29	4. Which number is the largest? ● 28

Page 75

26; 30; 27; 29; 28

Page 76

Count and write the number in the box.
Color the set that has **more** crayons.

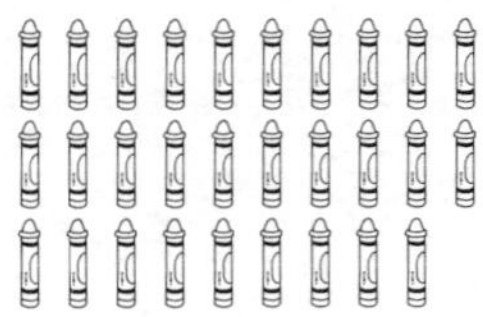

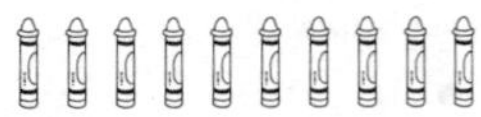

29

10

Color this set.

Count and write the number in the box.
Color the set that has **fewer** crayons.

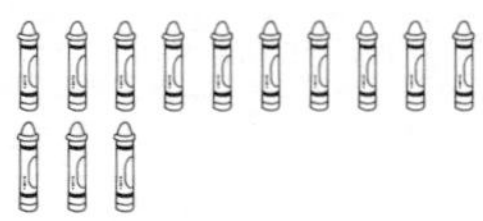

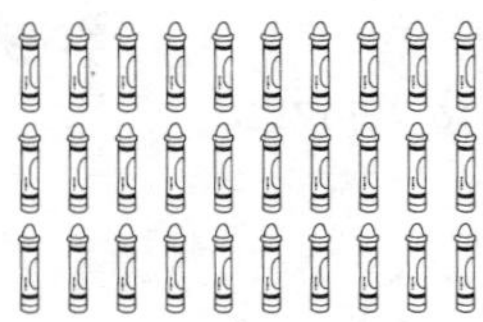

13

30

Color this set.